AF822000

Sustainable Building Design and its effects on Environmental Health

Dr. Aryan Amirkhani (Ph.D. in Architecture)
Tarbiat Modares University/Department of Architecture

Dr. Fatemeh. Tehrani (Ph.D. in Geography and Urban Planning)
Islamic Azad University, Science and Research Branch/Department of Geography and Urban Planning

DEDICATION

To Our Wonderful Readers

CONTENTS

ACKNOWLEDGMENTS

Thank you to everyone who strives to grow and help others grow

1 ABSTRACT

Evidence indicates that various building aspects influence the health and the comfort of occupants. Lack of safe drinking water, inadequate heating or cooling, waste disposal, and ventilation systems result in adverse health effects, including respiratory illness, asthma, infectious disease, injuries, and mental health disorders. These conditions, which are characteristic of substandard housing, predominantly affect vulnerable populations specifically people of radical minorities. [Krieger, Higgins, 2002] Moreover, susceptibility of poor and minority populations to hazard may be increased due to underlying health conditions, [Frumkin, 2002] The well- being, health, and safety of a building occupants and all those potentially affected by a building is a primary goal of a sustainable architecture. Health goals, on the other hand, relate to all issues dealing with the long term health of the occupants and passers-by. As the environmental impact of buildings becomes more apparent, a growing field called sustainable design is leading the way to reduce that impact at the source. Sustainable building is in fact creating environmentally sensitive, energy conscious and healthier developments. Generally one of the important points of green buildings is called "Quality of life", that means improving our living space through reduced indoor air pollution, smarter designs and improving environmental conditions for future generations. Since quality of life is essential part of green buildings, it is important to understand the relationship between technological sustainable development in different buildings and the impacts on the occupants and buildings owners.. In this paper social and psychological issues in green buildings are investigated in addition to applied behavior of occupants in these structures. This paper also investigates how indoor environment greatly affects the occupant's well being. The solutions to sustainable indoor environment are reviewed afterward. The results show that behavioural aspects of sustainability and how people interact with these structures can be important in sustainable design. The results also show that maintaining the conditions of a building improves the health of its occupants.

Index Terms—Sustainable, buildings, Occupants, Health, Effects, Comfort

2 INTRODUCTION

The growing use of synthetic materials, solvents and mechanical systems of environmental control within buildings carry threats to health and personal well-being. The term SBS is used to describe situations in which building occupants experience acute health and comfort effects that appear to be linked to time spent in a building, but no specific illness or cause can be identified. [Boxer, 1990] The complaints may be localized in a particular room or zone, or may be widespread throughout the building. SBS is generally seen as a problem caused by a combination of factors such as poor thermal, visual and aural comfort condition, the presence of gaseous pollutants, microbiological contamination, dust and fibers, and tobacco smoke. SBS is exacerbated when occupants of buildings lack the ability to perceive natural conditions and are unable to control their own internal environmental conditions. On the other hand, the term "building related illness" (BRI) is used when symptoms of diagnosable illness are identified and can be attributed directly to airborne building contaminants. SBS and BRI are associated with acute or immediate health problems.

In 1983, World Health Organization Committee reported that up to 30 percent of new and remodeled buildings worldwide may be the subject of excessive complains related to indoor air quality (IAQ). [World Health Organization, 1983] This condition may have long-term problems in some buildings. Frequently, problems result when a building is operated or maintained in a manner that is inconsistent with its original design or prescribed operating procedures, then indoor air problems are a result of poor building design or occupant activities. It is important to note that complaints may result from other causes. These may include an illness contracted outside the building, acute sensitivity, job related stress or dissatisfaction, and other psychological factors. Nevertheless, studies show that symptoms may be caused or exacerbated by indoor air quality problems. [Galea et al, 2005] Effective control of SBS requires an appreciation of the toxicology of materials, finishes and decorative veneers employed in building, furnishings and furniture. In addition, the quality of the internal environment may be affected by external pollutants finding their way inside such as traffic fumes, radon and landfill gases.

As much indoor air quality control is dependent upon using outside air for dilution and displacement of pollutants, the interior environmental quality relies upon clean exterior air. In some town centers, the outside air may be more contaminated than that on the inside. Often SBS problems are dependent upon a combination of factors and the mixing of pollutants and low-level toxicity from many sources. As psychological factors play their part, it has been observed that sealed air-conditioned buildings are more prone to complaints of SBS than buildings ventilated and lit by natural means. [Hedge et al, 1996] Since people spend the majority of their lives indoor the quality of the internal environment greatly affects their mental and physical health. The emergence of deep planned, relatively small windowed office buildings and shops in the 1970s onwards coincided with a growth in health problems. As energy conservation has become more important, buildings have tended to become more highly sealed to control unwanted air filtration.

With reduced ventilation rates, the low level toxicity inside become more critical to the well being of building occupants. The period 1970-98 has seen a big increase in synthetic carpeting, the use of adhesives, plywood and particle boards which are known to give off gaseous pollutants. At the same time, the growth in service industries led to greater use of office machines of various kinds which became another source of background pollution through dust and electromagnetic radiation. [Benefits of Good IAQ, 2008] These changes to the office environment have been paralleled in the home where wall-to-wall carpeting (usually incorporating man-made fibers), sealed window systems and the replacement of open fires by central heating, has led to concern over SBS in the domestic sector.

This background provides the clues to how a healthier internal environment may be created. However for the designers, the rate at which pollution is given off by certain materials is not readily available. The lack of necessary judgmental data means that well-intentioned architects remain in the dark. The concentration of air-borne pollutants from inside buildings may be a hundred times that in the outside air but little is known of their precise origin, whether in paints, cleaning solvent, particle board, pesticide applications, halogens and so on. Hence the task of designing SBS free buildings is particularly difficult. [Hedge et al, 1992] Some pollutants are certainly the result of design decisions, others stem from the nature of maintenance or management contracts, and others are introduced by the occupants themselves.

Insert chapter two text here. Insert chapter two text here. Insert chapter two text here. Insert chapter two text here. Insert chapter two text here. Insert chapter two text here. Insert chapter two text here. Insert chapter two text here. Insert chapter two text here. Insert chapter two text here. Insert chapter two text here. Insert chapter two text here. Insert chapter two text here. Insert chapter two text here. Insert chapter two text here. Insert chapter two text here. Insert chapter two text here. Insert chapter two text here. Insert chapter two text here.

3 INDOOR ENVIRONMENTAL QUALITY IMPROVEMENT

Indoor environmental quality improvement

As SBS and BRI create major physical and mental problems on occupants, it is significant to improve indoor air in buildings. To solve these problems, following suggestions may be appropriate:

1- Increasing ventilation rates: It can be a cost effective means of reducing indoor pollutant levels. HVAC systems should be designed, at a minimum, to meet ventilation standards in local building codes; however, many systems are not operated or maintained to ensure that these design ventilation rates are provided. In many buildings, IAQ can be improved by operating the HVAC system to at least its design standard. [Interpretations to ASHRAE Standard, 1989]When there are strong pollutant sources, local exhaust ventilation may be appropriate to exhaust contaminated air directly from the building. Local exhaust ventilation is particularly recommended to remove pollutants that accumulate in specific areas such as rest rooms, copy rooms, and printing facilities.

2- Pollutant source removal or modification: It is an effective approach to resolving an IAQ problem when sources are known and control is feasible. Examples include routine maintenance of HVAC systems, such as periodic cleaning or replacement of filters; replacement of water-stained ceiling tile and carpeting; institution of smoking restrictions; venting contaminant source emissions to the outdoors; storage and use of paints, adhesives, solvents, and pesticides in well ventilated areas, and use of these pollutant sources during periods of non-occupancy; and allowing time for building materials in new or remodeled areas to off-gas pollutants before occupancy.

3-Air cleaning: It can be a useful adjunct to source control and ventilation but has certain limitations. Particle control devices such as the typical furnace filter are inexpensive but do not effectively capture small particles; high performance air filters capture the smaller, respirable particles but are relatively expensive to install and operate. Mechanical filters do not remove gaseous pollutants. Some specific gaseous pollutants may be removed by adsorbent beds, but these devices can be expensive and require frequent replacement of the adsorbent material. In sum, air cleaners can be useful, but have limited application.

4-Flexible design: Designed to give occupants control over their interior environment. For reducing sick building syndrome in buildings and workstations, the building design could easily be modified in an adaptable condition, In other words, the building design should encourage tenant's motivation to control his or her own environment and provide a comfortable condition.

5- Education and communication: They are important elements in both remedial and preventive indoor air quality management programs. When building occupants, management, and maintenance personnel fully communicate and understand the causes and consequences of IAQ problems, they can work more effectively together to prevent problems from occurring, or to solve them if they do. [World Health Organization, 1983]

6- The location of building: Designers should pay attention to the appropriate site location. Buildings should be located where external air quality is high.

Psychological impacts of sustainable structures

About a decade ago the term sick building (SBS) was coined to describe the condition where people became ill simply by occupying a particular building. (Edwards, 1999) shows some effects of such buildings on human life in addition to some probable causes and solutions.

The symptoms of SBS (headache, nausea, stress, sore throats, asthma attacks, etc.) can cause genius distress in some, and mild discomfort in others. SBS is generally seen as a problem caused by combination of factors such as poor thermal, visual and aural comfort condition. SBS is exacerbated when occupants of buildings lack the ability to perceive natural conditions (daylight) and are unable to control their own internal environmental conditions. Many studies have shown the significant effect of nature on human life. Haber (Haber, 1977) founded that primary concerns of occupants in tall buildings were the great sense of disconnection from the nature.

As he demonstrated the separation from the nature in large urban projects and tall buildings may have significant negative effects on human life and may threaten social health in different aspects [4]. Kaplan's (Kaplan, 1995) attention restoration theory (ART) posits that natural settings provide an easy fascination that can help people recover from the mental fatigue caused by the more effortful attention often required by work . Wilson (Wilson, 1984) addresses that humans have an innate need for contact with the natural world and in particular seek to occupy and recreate in settings that are similar to those from our evolutionary past. Ulrich (Ulrich, 1993) reviews studies supporting hypotheses that humans have innate negative and positive associations with various elements of the natural world.

Nowadays psychologist found that that spending time in natural environment can have positive psychological effects on human mind (Hartig, Mang, Evans, 1991) Kuo and Sullivan (Kuo, Sullivan, 2001) have extended these findings by showing that the presence of vegetation in public housing sites is significantly associated with reduced levels of aggressiveness and crime. They even show that residents living in “greener” surroundings report lower level of fear, fewer incivilities, and less aggressive and violent behavior. (Kuo, Sullivan, 2001) The hypothesis about the restorative function of natural environment has been tested in many empirical studies. Ulrich (Ulrich, 1994) for example founded that hospital patients who could look out on trees and nature from their windows recovered more quickly than whose views restricted to buildings. Schroeder (Schroeder, 1991) has shown that natural environments with vegetation and water induce relaxed and less stressful states in observers compared with urban scenes with no vegetation. This ability of natural elements to function as "natural tranquillizers" may be particularly beneficial in urban areas where stress is an all too common aspect of daily living (Berg, van den Top, Kranendonk, 1998).

Later studies have lead to similar results, strengthening the assumption that natural environment have positive influence on psychological and mental health. Not only green areas have significant effect on human behavior, but also natural features seem to affect emotional well being. For example an evaluation of day lighting in schools showed that students in day lit classrooms were happier than those with minimal day lighting in other classrooms. (Heschong, Wright, Okura, 2002). Improving air quality can also create healthy life. Fisk et al. (Fisk, Price, Faulkner, Sullivan, Dibartolomeo, 2002) suggested that 16–37 million cases of colds and flu could be avoided by improving indoor environmental quality, resulting in $6–$14 billion annual savings in the United States.

The symptoms of SBS might also be reduced 20–50%, resulting in $10–$30 billion annual savings in the United States (Milton, Glencross, Walters, 2000). Similarly, people seem to experience more headaches, dizziness, and tiredness when toxic materials are present (Wargocki, Wyon, Baik, Clausen, Fanger, 1999). Therefore, occupant health is greatly related to the natural features. Generally green buildings that provide natural day lighting, appropriate ventilation, use of none toxic materials, natural landscape views and vegetation inside the building can improve human health and have positive psychological effect on occupant's mind. (Wise, Betch, 1999)

Occupants` behaviors in sustainable buildings

Designing a successful green building, architects should predict what human activities are required more than others; also they should use this knowledge to design facilities to support their predictions. It is important that different places require various facilities. For example an industrial factory must provide enough space for manufacturing equipment and processes, attending to acoustic and lighting needs, facilities for employee meetings, breaks, and meals.

If the space is to be used as a hotel, it must support different room size, lightning requirements, public lobby, waiting hall, management room, good ventilation and enough fresh air. In 1996 the problem of stress and eye strain associated with VDUs and other office machines was very serious. To solve the problem, the EC introduced Directives "Display Screen Equipment" and "Workplace" (Edwards, 1999).The Directives address the balance between general internal environmental standards and the specific standards at the workstation. In the process they put the control of the lighting environment more in the hands of the office worker by requiring the installation of blinds, curtains and individual lamps. A Gallup report in 1990 "Productivity Loss in the office" (Beer, 1994) found that companies are losing 15 percent of productivity every year due to light-induced stress and illness. This amounts to more than $2500 per employee per year, a sum more than adequate for the cost of adapting most working environments to the new standards. There are dangers in the directives from an energy point of view.

The need to fit blinds on every office window could lead to higher levels of artificial lighting in the core of the building. As lighting in offices is often the major source of heat, this will add the pressure to mechanically ventilate the building (with increased energy costs). A naturally ventilated office with opening windows is to some extent undermined by having to pull blinds and curtains. In addition, the screening of daylight at windows will inevitably block views with the possibility of adverse psychological effects.

The energy consumption in buildings is also related to construction design methods. Hence, an important concerning factor about occupant's action in green buildings is that architects should estimate how much energy and water will be consumed, when determining the required technical performance.

Occupants make behavioral choices that affect these systems, such as when to increase their personal comfort by adjusting thermostats, opening or closing windows, lowering or raising blinds and drapes. These occupants' choices are critical in determining how well and efficiently building system performs. In some countries traditional green buildings are typically planned to be efficient in these respects, but there are still not enough anecdotal experiences or formal evaluations of sustainable modern buildings in order to determine how effectively they respond to variations in human needs.

In 1998 Heerwagen and Wise (Heerwagen, Wise, 1998) studied about the benefits of green buildings and building performance to occupants' behavior as well, they addressed a green factory in Michigan through their studies. This factory was found to be operating sub-optimally in several respects. For example, energy consumption was higher than expected, and workers kept large bay doors open for the fresh air and for the beautiful landscape that these doors provided. While the original design had allowed for better ventilation and good views, high shelving was added after initial occupancy that blocked the vents and views, encouraging occupants to open the bay doors. This study shows that all facilities in green buildings should be very sensitive to variety of user behavior and could suffer greatly if the actual usage varied significantly from that predicted in the designing stages. Generally if occupants were unable to conform to high standards, such as taking brief showers, turning down thermostats in the winter or buying minimally packaged products, the building's performance could fall below initial predictions.

Social - psychological studies and applied behavioral analysis are two significant approaches on conversation maters in psychological research. The former provide attempts to understand and change attitudes, and the latter, which describes critical behavioral contingencies. Logically, social-psychological studies are based on the intuitive notion that attitudes underlie behavior, and these two items are related to each other. It means that, by changing the former, the latter will change too. A brief review on social-psychological studies show there is often little correspondence between attitudes and behaviors and that in some cases attitudes follow behavior changes rather than preceding them (McKenzie, 2000).

Another point that might be effective on occupant's behavior is training and conservation education. Many studies have shown there is a lasting effect on tenant's attitude when some strategies elicit public commitment to a pro-environmental action (winter, 2000). Psychologists founded that there is a complex relationship between environmental attitudes and behaviors. In fact, the behaviors involve with a variety of personal and contextual factors such as social and demographic factors, incentives and constraints, knowledge, values and commitment. What is important is that changing any one element may not be sufficient to effect behavioral change. (Stern, 2000) Therefore some psychologist point out that it may be necessary to address specific situational barriers that inhibit change if they want to approach to effective policy change. For example, McKenzie-Mohr through his studies about sustainable behavior in community indicated a combination of psychological approaches and marketing strategies to reduce water consumption. (McKenzie, 2000)

For approaching to a clear conservation behavior, scientist divided the applied behavioral analysis into 2 parts:

1- Events that precede and cue or prompt the actual target behavior, called antecedents.

2- Reinforcing or punishing results of a behavior that determines its likelihood of being repeated, called consequences. (Geller, Winnett, Everett, 1982)

Applied behavioral analysis of energy use by consumers suggests that the nature of both the antecedent and consequent conditions in most buildings often consume greater energy use while discouraging conservation behavior (Geller, Winnett, Everett, 1982). What is significant in behavior theory is that the consequences of a behavior need to be urgent in order to be effective. Nearly all the events that might discourage energy use appear after a significant and very long delay. For example, the cost of energy, which is typically presented in the form of a bill arriving days or weeks after the thermostat, can constrain a tenant from turning up the level of air conditioning in the summer or the heat in the winter. If there are enough reminders to focus tenant's attention on the cost of energy and courage them to conserve it, a great amount of energy will save. Unfortunately, there are still few obvious signals in most buildings that might serve to remind consumers of the costs of energy and prompt them to conserve.

Changing the occupant's behaviors and actions can provide prompts for conservation at the behavioral decision point. For example, Becker and Seligman (Becker, Seligman, 1978) placed a light in the kitchen that went on when the air conditioner was operating but outside temperatures were cool. Others have offered residents daily written information on the amount of electricity used or installed prominently placed and easy to read meters that show energy costs in real-time (Kohlenberg, Phillips, Proctor, 1976). Evaluations have suggested that these techniques may result in a significant reduction of overall energy use or during peak time use (Geller, Winnett, Everett, 1982). Sometimes consumer behavior and its consequences are particularly disconnected such as in large buildings where those who pay the energy bills are not the ones controlling energy use.

The environment refers to a set of physical elements, social and economic conditions, and generally refers to phenomena and events that involve man or are organized by him. Today, the role and importance of the environment in the formation and occurrence of human behaviors do not need to be demonstrated. To better utilize capital and resources, all aspects and dimensions of environmental design should be considered, including users' needs and conditions. Designing, planning, and adapting the environment should look for its role and responsibilities in deeper and wider areas than human beings, life and complex relationships within it. In the relations between humans and the surrounding environment, sustainable development has always been a fundamental concept. By defining sustainable development as development of quality of life and life in the capacity to tolerate backup ecosystems, and with regard to concepts such as development and the environment, the relationship between human, environment and sustainable development is explained in the context of this definition. Various causes play a role in the formation and sustainable development of urban environments, among which climatic considerations play an important role. Climate change in urban environments, including natural reactions to current interactions in urban environments, can be considered in two dimensions:

1. Climate as one of the components of ecological systems.

2. Climate as a natural and open system

Climate as one of the subsystems involved in the ecological production of nature with other components of the system is in mutual interaction, and therefore responds to changes occurring in natural systems, and adapts itself to new conditions.

Therefore, any change in the system of nature in some ways leads to the occurrence of climatic changes. In the next step, the climatic factor can be considered an independent natural system (relative independent meaning). The climate system is also an open and dynamic system, and for this reason, every change is made in other components of the climate system. Whenever enough attention is paid to the last two points, the complexity of the explanation of climatic variation is self-evident. At the same time, it should be noted that within the framework of urban studies, as a result of this study, only climate change is considered, which occurs at the maximum on the mesoclimate scale. Nevertheless, we also know that macroclimate, under the influence of ozone layer erosion and increasing the thickness of the greenhouse layer of Earth envelope, is a profound climatic development that undoubtedly affects the microclimates; as a result, the urban climate will be affected by its consequences.Within the framework of this discussion, the following climatic changes are briefly discussed. In fact, these are the changes that, in the sense of "urban climate", offer a special identity:1. Irradiance, transparency and / or air turbidity; 2. Air temperature; 3. Air flow or wind system; 4. Humidity; 5. Quantity and quality of atmospheric precipitation; 6. Air pollution

After mentioning the factors affecting the urban climate, climatic solutions and considerations in urban spaces can be briefly analyzed in terms of performance in places such as walking paths, riding paths, and green spaces. Obviously, these considerations are factors influencing urban environments. At the end, the results of the above discussion are expressed.

Different concepts and dimensions of sustainable urban development

Ambiguity in the sense of sustainability has led to a kind of uncertainty in the concept of sustainable urban development. There are many researches to overcome these doubts, but they are not still sufficient. The city is a complex system, and as the larger it is, the result of development measures will be more unpredictable. However, the two concepts "urban sustainability" and "sustainable urban development", which are often used owing to their proximity to each other, should not be mistaken with each other.

To distinguish these two, we must note that sustainable urban development, in fact, represents the process during which stability can occur, but stability is a set of situations that persists over time. However, some consider sustainable development in utilizing land and encouraging the reuse of buildings, and maintain that the size, density and location of human habitats suitable for sustainability, will be variable in line with the development of technology in energy, production, building, and transportation. Some also define sustainable development as a way of improving the quality of life and life in the capacity of bearing backup ecosystems. In each form, various factors and dimensions are effective in sustainable urban development; climate and environmental considerations are only one of them. Table 1 summarizes the importance of sustainable urban development.

General considerations in planning for sustainable urban development	1. Economic considerations
	2. Social and cultural considerations
	3. Physical considerations and principles
	4. Management, decision-making and executive practices
	5. Climatic and environmental considerations

Table 1: Important dimensions for sustainable urban development

Climatic and environmental considerations

In the following, some of the main factors affecting climate change are mentioned, which are important in the study of urban environments and planning for sustainable urban development.

Radiation and air turbidity

A commonplace reality, which has disappeared from the realm of consciousness of man is that the energy needed to maintain and sustain life on the planet is received from the sun. However, if the planet was to receive only solar energy, it would have to increase its temperature regularly. Nevertheless, since such a situation does not occur in natural conditions, it must be admitted that the earth also reflects energy to the atmosphere. It is precisely for this reason that the energy received and the reflected energy must be assumed to be a kind of equilibrium. Research in recent years has indicated that the balance between two energy streams is disrupted, which is why the temperature of the earth is increasing. In natural conditions, to establish balance between the received energy and reflected energy, there are a number of very complex and sensitive processes between the sun, the space above the atmosphere, the atmosphere and the earth.

The emergence of climate change on a global scale is an indication of the disruption of the natural process of these processes. In urban environments, air pollution causes changes in the way solar radiation enters the earth's surface. As air pollution increases, the amount of solar radiation received by the earth is reduced, and the amount of reflection by the polluted layer over the city is increased. In this regard, the amount of suspended matter and aerosols in the city has a decisive role. When there is polluted air over the city, it prevents ultra violet short wave transmission (less than 0.36 micron) and reduces its input. The visible light spectrum along the ultraviolet spectrum is weakened by passing the polluted atmosphere. Under the influence of suspended particles in the air, direct light is dispersed and the ratio of these two to each other differs in favor of diffuse light. Reduction of the clarity of the lines and edges as well as decrease of the range of vision are among the most significant consequences of these changes. In the distribution of sunlight, the horizontal and vertical arrangement of the city also interferes. In the streets surrounded by high-rise buildings, the roofs of the buildings become the main source of energy exchange. The sidewalks and the lower parts of the buildings mainly receive reflections from the roofs and walls.

Changes in urban air temperature

The emergence of changes in air temperature in urban environments should be considered in several ways: increasing thermal energy discharge in the air, changes emerging in thermal energy distribution, increasing the city's capacity for thermal storage. A massive amount of thermal energy is daily discharged from the combustion of fossil fuels (home supplies, motor vehicles, industrial resources), and this amount of energy is steadily increasing along with population growth and industrial activities. Changes in the natural system will result in climate change. Air pollution reduces the clarity of the lines and the vision range. The use of old and outdated techniques for the use of fossil fuels not only increases the pollution of the air, but also exacerbates its consequences. In a situation where the condition of the city's airspace - whether for natural reasons or because of the inappropriate design of the city's macroform - is disturbed, this thermal energy, which is abnormally discharged into the environment, can have profound changes in the climatic structure of the city. In urban environments, thermal distribution also changes strongly due to the change in the level of thermal exchange resulting from urban constructions.

Regarding the different levels that a building can have, the levels of its thermal energy received, in comparison to the land surface occupied by the building, are multiplied by several times, and the same amount of received and reflected thermal energy is multiplied. In other words, increased levels of construction and the multiplicity of buildings in the city have a direct impact on the production and consumption of solar energy in the city. In addition to increasing the storage capacity of thermal energy, a kind of movement in the process of increasing and decreasing the temperature of the city also occurs. In the warm months, the heat storage in the building's structure makes it possible to delay the cooling of the city during the night, since at this time, the walls, ceilings, paving blocks discharge the thermal asphalt levels stored during the day, thereby delaying the natural cooling process of the environment. The phenomenon is directly related to the size of the city and the multiplicity of buildings. The city's thermal capacity especially increases when the city is poor in terms of heat-absorbing spaces and surfaces (urban forests, extensive trees, lakes and wetlands).

Changes in air flow and wind system

The wind system can be classified into two major groups:

Regional winds that have developed in the macroclimate range and in different months and seasons, are influenced by a pattern that is more or less specific to a particular space.

Local winds that appear in Dashtkouh and Kouhdasht flows in most of the mountain cities of Iran. In the coastal areas of the north and south of the country, they are also among the typical local winds.

Certainly, in many other parts of Iran, there are other local winds that mentioning them one by one is outside the current discussion framework. The important point here is that the macroform, as well as the arrangement of the spatial organization of the city, can have a serious impact on the natural process of winds and, consequently, the role that winds can play in ventilating and purifying the city's air. Here, the role of the city's size, the multiplicity of buildings, and the design and organization of the height of the city's buildings are well revealed. Increasing friction within the city limits, makes the movement of the air difficult, especially to a height of 1000 meters from the surface of the earth.

Part of wind energy is used to counteract frictional forces when passing through and over long, short-lived structures, and therefore its speed is reduced. Furthermore, when the air flows into the city, a "pillow from the air" is created on the city, which is followed by the flow of air. The air flow that moves on this pillow cannot penetrate the surface of the earth and help to ventilate the city's air. The situation intensifies when the air corridors to the city are blocked by high-rise buildings. In the streets, the wind conditions vary, depending on how the spatial arrangement around the street borders, the green spaces, and the wide, long or short buildings, changes. The street that are built in the direction of the prevailing winds and are surrounded by long and narrow buildings usually increase wind speeds. The joint effect of urban construction, thermal island, and wind flow is also significant in urban environments.

Thermal islands are usually formed in the densely populated areas of the city with poor thermal discharge. In this dense atmosphere, the air close to the ground surface is warm and light, and climbs upwards. In this case, a low pressure center is formed which can cause air pollution around the city. Such a phenomenon can have both positive and negative aspects. In a case where the city is surrounded by extensive forest parks and water levels, the air that is blown into the city is a clean, oxygen-rich air and, therefore, can help to ventilate and increase the city's air quality. Moreover, when the city has been occupied by industrial and polluting air around the city, the air entering the city is mainly polluted air, which can ultimately affect the air pollution of the city. The above-mentioned situation can be considered absolutely true in the case of Tehran, and relatively true in the case of Isfahan. In the case of Tehran, from west to southwest and no local winds of Dashtkouh, none of them can considerably contribute to the air conditioning of the city. At the same time, it should be noted that, in terms of air purification, only when the wind speed reaches more than 3 m/s, it can play an effective role. However, in Tehran, due to its unorganized macroform, as well as the natural weakness of the winds, the air conditioning capacity of the city by this climatic factor is very low.

Changes in air humidity

Identification of air humidity in urban environments, like air temperature, has its own complexity. This complexity in the cities of Iran, which is in the Iranian Turanian plant territory from the perspective of plant geography (the vast majority of Iran, with the exception of the north biome of the Alborz mountain ranges and the biome under the influence of the Persian Gulf and the Oman Sea), becomes more complicated for several reasons:

- This plant area, in its natural range in its plains, is formed of mostly tight trees land, bush, shrubland, semi-desert and desert steppes. The characteristics of these areas are long-term drought, and therefore numerous bioclimatic constraints and very limited ecological vulnerability. When a place from this plant kingdom turns into a city, the ecological construction of the natural environment generally changes. The outcome of these changes, which will be discussed below, brings the urban climate to the point of drought more and more.

- On the other hand, the creation of urban green spaces with irrigation, the creation of agricultural lands around or near the city, the creation of artificial forests to deal with the sand, the massive use of fossil fuels in the city increase air humidity. Thus, as seen, in urban environments, there are two completely opposite flows that move the environment toward the drought and the other to increase the humidity of the air. In the past, some changes have been made, which will increase the temperature of the city. In such a situation, it is obvious that the city's air is becoming dry. Nevertheless, this hypothesis does not always turn into reality.

The existence of private, semi-private and public green spaces, which in Iran's dry environments always require irrigation, some moisture enters into the air by evaporation of plants, and as a result of water evaporation from the surface of the soil. It has also been mentioned that fossil fuels after combustion always enter some water vapor into the air. For example, as a result of combustion of one liter of gasoline in a vehicle, one cubic meter of vapor is produced into the air. The city of Tehran is a prominent example of this situation; in this city, it was expected that by 8 million liters of gasoline daily, 8 million cubic meters vapor would enter into the air. In addition, gas produced from gas, gasoline, oil and mazut should also be considered. At the same time, as the traditional architectural style has changed into a so-called modern architecture, the use of cooler has become essential in many buildings. On the roofs, air conditioners are exposed to intense sunlight on the one hand, and on the other hand, due to technical defects, in most cases, there is a significant loss of water. These two factors, each with their own share, can enter some moisture into the air.

The appearance of changes in atmospheric precipitation

Climate in the concept of the weather pattern in a given time and place is the product of the joint operation of various meteorological factors and therefore follows the rules governing the open and dynamics systems. Therefore, as previously mentioned, if one or more parts of the system are changed, the other components will change in order to adapt to the new state of the system itself. Therefore, due to the climate change mentioned earlier, it can be expected that the trend of rainfall in urban environments will also change. This climate change also applies to climate as a subsystem of the natural system and the climate as an independent natural phenomenon. The most obvious form of these changes is the reduction of precipitation in the form of snow; this phenomenon, as a manifestation that has occurred over the last few decades, has even played a role in the memory of most citizens. Most scholarly sources emphasize the fact that under conditions of decreasing precipitation, snow is added to the amount of rainfall. However, due to lack of sufficient information on the climatic condition of Iranian cities, this claim cannot be commented on the validity of this claim. In addition, the frequency of lightning and heavy rainfall are also attributed to urban climate. Furthermore, the frequency of the occurrence of convective rains or rains arising from the rapid movement of hot air based on the city is also considered the characteristic of urban climate.

When the hot air, mounted on a city in a state of high instability, climbs upwards, this displacement and rapid drop in temperature can lead to convective rainfalls. A typical example of this phenomenon was the black rain after the fire of the mazut pools of Tehran Refinery in April 24, 1986. In those conditions, the massive amount of heat energy discharged in the air climbed upwards and quickly climbed into the upper layers of the air, causing rainfall. At the same time, since the amount of soot was discharged through the burning mazut, the soot was washed by rain and returned to the earth as black rain. In the winter, in urban environments, plenty of weathering and air pollution are also higher than in open and non-built spaces around the city. In this regard, two main reasons can be mentioned:

Water vapor, which is discharged into the cold air in different ways

Particles of air pollutants acting as primary condensation nuclei and providing conditions for the accumulation of water vapor around them.

Climate change caused by air pollution

The above-mentioned climatic changes are directly or indirectly to air pollution. Reduced transparency or increased air turbidity is in direct contact with or increased air turbidity is in direct contact with air pollution by suspended particles, vapors, and chemical compositions of various pollutants in the air (mist smoke of London or Los Angeles). Part of the increase in the temperature of the city should also be taken into account for the city's air pollution. Air pollution also affects changes in humidity and atmospheric precipitation.

The effect of environmental factors on pedestrians as one of the main users of urban space is studied. Then, how sustainable design in urban spaces is functional in terms of designing sidewalks, riding paths, green spaces, and urban furniture.

Impact of environmental factors on pedestrians

Environmental factors have a significant effect on the residents of the neighborhoods and their willingness to participate in open-air residential areas. The climate is a limiting factor for using open space in very hot or very cold days of the year. Fresh air, suitable temperatures, favorable relative humidity, relatively gentle air flow and protection against rain provide comfortable conditions for humans. All climatic factors such as air temperature, air humidity, solar radiation, wind, rain, and freezing affect pedestrians. By referring to the standard (Penwarden) (Graph 1), it turns out that the temperature range of 16 ° C to 21 ° C is very comfortable for pedestrians.

All climatic factors such as air temperature, air humidity, solar radiation, wind, rain, and freezing affect pedestrians. By referring to the standard, it turns out that the temperature range of 16°C to 21°C is very comfortable for pedestrians. At temperatures of 4.4°C to 12°C, the existence of sunshine in thoroughfares is possible, and at -25°C, the air is windy and can cause the air to cool on the sidewalk.

Climate and topography are among other factors affecting pedestrians. Water can have heat and cold effects. The greater the amount of water, the more impact it will have on microclimate. Green space also has a considerable influence on microclimate. The temperature of parks is higher at night and lower during the day than the outside space. One of the other climatic factors is wind that can be controlled with vegetation or structural cover.

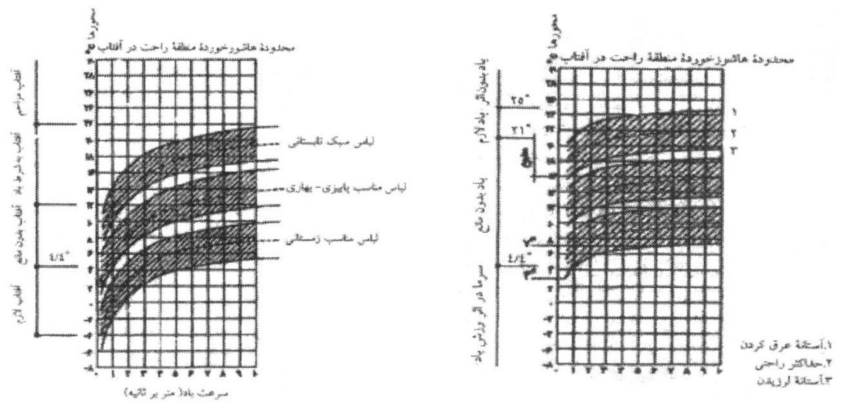

Graph 1: Comfortable thermal status in terms of wind speed and type of cover for pedestrians according to the standard of Penwarden

Components and elements in urban areas

Sidewalks: The factors to be considered in the design of pedestrian routes are as follows:

- Connection of pedestrian routes will be achieved by suitably designing and separating pedestrian path from crossings, nonuse of motor vehicles and bicycles by pedestrians, non-occupancy of part of the sidewalk by urban or construction equipment, proper footing of the sidewalk, and not cutting the paths by bridges or natural obstacles.

- Short access paths, because passersby usually chooses the shortest route.

- Safety of sidewalk routes

- The comfort of the footpaths, some of which are: proper slope, smoothness and stability, lack of slippery footing, adequate width of the sidewalk with regard to pedestrian traffic, comfortable passage of physical and motor skills, and gathering surface water of sidewalk.

Green space: Green space is part of the urban open space, where its natural or artificial arenas are under the deployment of trees, shrubs, flowers, grasses and other plants. It is also maintained or created to improve the living, habitat and welfare of the citizens based on human monitoring and management by considering its related rules and specialties. The most important environmental effects of green space in cities are their environmental functions, which include reducing air pollution, reducing noise pollution, producing oxygen and carbon dioxide, controlling radiation and reflecting light, controlling wind, aesthetics, creating shadows and landscapes, and air refinement. The green space, when placed in the urban body, finds the tendency of the urban hierarchy and is divided into different scales in the city. Usually, how to use the green space is determined by the distance indicator and its location in accordance with the urban hierarchy.

Urban furniture: Open space consists of various details and components made by mankind, such as sitting places, signs and guides, furniture, and lights.

Sitting places: Sitting is influenced by two specific conditions: climate and space. In short, sitting should only occur where the external environment conditions are desirable and suitable for sitting places that are more accurately selected than standup sites. The considerations that should be taken into consideration when designing sitting areas in open spaces are: Bench type, location of bench placement with regard to the sighted views, types of sitting places and special resting benches for the elderly and people with physical-motor weaknesses. Undoubtedly, observance of the above principles and efforts to control the climate change affecting urban environments are all factors that lead us to sustainable development and sustainable cities.

D. Conclusion

The issue of sustainable urban development has been considered an important scientific issue in all communities in recent years and has received a large part of the literature on urban development. In this regard, different dimensions of sustainable development from different angles have been investigated and evaluated, and various approaches have been considered in different perspectives. A comprehensive approach to urban development policies and plans, as well as formulation of integrated economic, social and environmental objectives, are at the forefront of policymaking and sustainable urban development planning. Establishment of a balance between large and small urban centers, cooperation between private and public sectors, as well as development of a local strategy are among sustainable urban development policies to be considered. Among various aspects affecting sustainable urban development, climatic factors are important as an influential element. Observing the practical solutions and considerations and attempting to create a balance in the urban environment, in accordance with the climate change, are one of the issues that leads us to achieve sustainable development.

separation from natural elements, often occurs in large buildings, has negative physical and psychological effects on human life. Since there is an increasing attention paid to larger building projects, architects should pay more attention to occupant's health on buildings. This article contributes to a growing dialogue addressing the impacts of sick buildings syndromes and poor indoor environmental quality on human health, by highlighting sustainable strategies that effects the indoor environment and concurrently improve occupants' health. Moreover understanding the appropriate ways attending to psychological needs can improve the success of meeting the sustainability goals. Therefore, designers should attend to the strategies that encourage occupants to observe sustainable behaviors.

References

[1] Berg, A., van den Top, I.M., Kranendonk, R.B, Natuur wensen van stadsmensen (The Demands on Nature of Urban Citizens) (in Dutch). IBN-DLO Rapport, Wageningen, the Netherlands, 1998.

[2] Becker LJ, Seligman C. Reducing air conditioning waste by signaling it is cool outside. Pers Soc Psychol Bull; 4:411–7, 1978.

[3] Board Niir, Modern technology of waste management: pollution control, recycling, treatment & utilization, Publisher Asia Pacific Business, 2000.

[4] Case FD. Design implications of living with passive solar homes. J Archit Plan Res;1(2):84–105, 1984.

[5] Edwards Brain. Sustainable Architecture(European directives and building design) , Second edition, Architectural Press, London, England,1999

[6] Fisk W, Price P, Faulkner D, Sullivan D, Dibartolomeo D. Worker performance and ventilation: analyses of timeseries data for a group of call-center workers Paper presented at the indoor air, 2002.

[7] Geller ES, Winnett RA, Everett P. Preserving the environment: new strategies for behavior change. New York: Pergamon; 1982.

[8] Haber GM. The impact of tall buildings on users and neighbors. In: Conway D, editor. Human response to tall building. Stroudsburg, PA: Dowden, Hutchinson, & Ross, Inc; p 45–55,1977.

[9] Hartig T, Mang M, Evans GW. Restorative effects of natural environment experiences. Environ Behav, 23(1): 2–25, 1991.

[10] U. Duncombe, "Infrared navigation—Part I: An assessment of feasibility," IEEE Trans. Electron Devices, vol. ED-11, pp. 34-39, Jan. 1959. Bahreini, Seyyed Hossein & Mitra Ahmadi Torshizi, "Study of Biodiversity Basis in Urban Structure", Journal of Environmental Studies, Vol. 23, Summer 1999, 35, Faculty of Environment, University of Tehran.

[11] Salami, Reza, 1997, "A Review of the Perspective of Sustainable Development, Rahyaft Journal, No. 17, Tehran.

[12] Radcliffe, Michael, 1994, Sustainable Development, Translation by Hossein Nayyer, Ministry of Agriculture Publications.

[13] Bulletin of the National Committee for Sustainable Development, Environmental Protection Agency, No. 3 and 2

[14] Bahreini, Seyyed Hossein, 1996, "Analysis of Urban Spaces in relation to Behavioral Patterns of Users and Design Criteria", University of Tehran, Tehran.
[15] Moore, Jean, Allen Iyo Mary, Mary Lees Sairy, 1994, "Urban Spaces, Design, Implementation, Management", Translation by Hassan Rezaie, Public Relations and International Affairs Office of Tehran City, Tehran.
[16] Razjuyan, Mahmoud, 1990, "Comfort by Climate Architecture", Shahid Beheshti University Press, Tehran.
[17] Planning and Budget Organization (formerly), 1996, Deputy Director of Technical Affairs, Office of Research and Technical Criteria, "Walking Facilities", Two Volumes, Publications of Socio-Economic Center, Tehran.
[18] Ministry of Housing and Urban Planning, 1988, Tehran Comprehensive Organization Plan, Volume 10, ATEC Consulting Engineers, Tehran.
[19] Ministry of Housing and Urban Development, 1996, Urban Planning Regulations (Access Pathways), Department of Urban Planning and Architecture, Tehran.
[20] Engwicht, David, 1992, Toward an eco-city, calming the traffic, envirobook, Sydney.
[21] Girardet, H, 1999, Creating sustainable cities, Green books, Dartington.
[22] Thayer, R, 1994, gray world, green heart, technology, nature and sustainable landscape, john willy and sons, Inc, New York.
[23] Van der Ryyn, 1996, Ecological design, Island press, Washington DC.
[24] McLaren, V.W, 1996, urban sustainability reporting, JAPA, vol62, Spring.

ABOUT THE AUTHORS

Dr. Aryan Amirkhani received his PhD in architecture from the University of Tarbiat Modares, Tehran, Iran. He has published over 150 national and International articles indexed in highly cited databases. He has been awarded some of the most authentic and credible Iranian academic positions as the top university researcher of the years 2009-2010, member of the Iranian national elite organization (national elite 2009-2010), the author of the best Iranian architecture book 2012-2013(honorable degree), etc.

Dr. Fatemeh Tehrani is a PhD holder of Geography and Urban Planning, in Islamic Azad University, Science and Research Branch, Tehran, Iran. She is the author of many different highly cited articles. She has working experiences as the head coordinator of different municipality affiliate companies and offices in Tehran.

Made in United States
Orlando, FL
04 November 2026

90207449R00021

ZEITSCHRIFT FÜR ANGEWANDTE GEOLOGIE

15 JAHRE
DEUTSCHE
DEMOKRATISCHE
REPUBLIK

HERAUSGEGEBEN VON
DEM ZENTRALEN GEOLOGISCHEN INSTITUT
UND DER ZENTRALEN VORRATSKOMMISSION
DER DEUTSCHEN DEMOKRATISCHEN REPUBLIK

AUS DEM INHALT

AKADEMIE-VERLAG · BERLIN

BAND 10 / HEFT 10
OKTOBER 1964
SEITE 509—566

Die ZEITSCHRIFT FÜR ANGEWANDTE GEOLOGIE berichtet ständig über folgende Arbeitsgebiete: Geologische Grundlagenforschung und Lagerstättenforschung / Methodik der geologischen Erkundung / Ökonomie und Planung der geologischen Erkundung / Technik der geologischen Erkundung / Geologie und Lagerstättenkunde im Ausland. In der Zeitschrift können alle strittigen Fragen der praktischen Geologie behandelt werden. Die Autoren übernehmen für ihre Aufsätze die übliche Verantwortung.